the forest

All inquiries should be addressed to:
Barron's Educational Series, Inc.
250 Wireless Boulevard
Hauppauge, New York 11788

Library of Congress Catalog Card No. 91-7743

International Standard Book No. 0-8120-4709-5

Library of Congress Cataloging-in-Publication Data
Sánchez, Isidro.
   [Bosque.   English]
   The forest / I. Sánchez, C. Peris — 1st ed.
       p.   cm. — (Discovering nature)
   Translation of: El bosque.
   Summary: Describes a trip to the forest, to learn about the trees,
animals, plants, and insects found there.
   ISBN 0-8120-4709-5
   1. Forest fauna—Juvenile literature.   2. Forest flora—Juvenile
literature.   [1. Forest animals.   2. Forest plants.] I. Peris, C. (Carme).
II. Title.   III. Series: Sánchez, I. (Isidro).   Discovering nature.
QH541.5.F6R5813   1991
574.5'2642—dc20

                    91-7743
                     CIP
                     AC

Legal Deposit: B. 14.943-91
Printed in Spain
1234   987654321

discovering nature

the forest

I. Sánchez

C. Peris

CHILDRENS PRESS CHOICE
A Barron's title selected for educational distribution
ISBN 0-516-08451-8

Today our class is taking a trip to the forest!

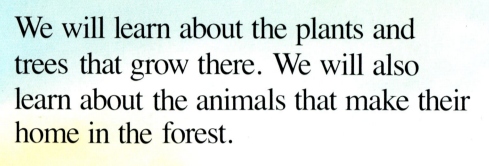

We will learn about the plants and trees that grow there. We will also learn about the animals that make their home in the forest.

Our teacher tells us, "This is a pine tree. The squirrel loves the tiny nuts that grow inside the pinecones."

"This tall tree is a spruce. Like other evergreens, it can be used as a Christmas tree."

"This big oak tree has a very thick trunk. Squirrels eat its fruit, which are called acorns."

"You can tell what tree a leaf is from by its color and shape. It is fun to collect leaves."

"The trees in the forest provide wood for paper and furniture, and many other things we need."

"We must be very careful in the forest. A single spark can start a terrible forest fire. The fires destroy the trees and harm the animals."

We like the forest. It is very beautiful and is home for many interesting birds, insects, animals, and plants.

Many different types of mushrooms also grow in the forest. But we must be careful not to take any, because they may be poisonous!

We sit in the forest and draw pictures of trees. Then we label every part of the tree—the leaves, branches, trunk, and bark.

We dig holes and plant little trees ourselves.
We tie them to stakes to keep them straight,
press the soil down, and water them well.

When they grow, we will be proud to have planted these trees and happy that we have added something to the beautiful forest!

THE FOREST

The forest, school of nature

A forest—even a small cluster of trees or a public park—is full of wonderful surprises that parents and teachers can introduce to children. They can be taught to discover a great variety of animals, plants, and insects that live in these areas, and learn how the lives of some of them depend on the others. This provides an educational experience that is both stimulating for the adults and delightful and informative for children.

Under the direction of the parent or teacher, the visit to the woods can be made a game of exploration. Thus, without losing any of the excitement that comes from enjoying a group activity, each child will be made to feel that he or she plays a big role in this great learning adventure! A well-planned outing can be the first step in encouraging children's interest in nature and teaching them how to respect it.

How many different trees are there?

The teacher can suggest many activities through which children can learn about the different trees in the forest. They should be taught to distinguish them by means of:

- the bark of their trunks,
- the shape of their leaves,
- the shape and composition of their fruit, and
- the overall shape of their trunk, branches, and (when present) the leaf mass.

What other plants are there in the forest?

Trees are only the tallest and most obvious plants in the forest. Plants that are lower to the ground and, therefore, within the children's reach, attract them with their fruit and flowers. The teacher can suggest some activities that will increase the children's curiosity. These games should consist first of only *looking* at different shrubs, flowers, and fruit. *Restrict gathering to the species that you know to be safe.*

Obviously, this warning must be stressed, especially where mushrooms are concerned. Children must know that some species are extremely dangerous and should not even be touched. The same warning applies to poison ivy, which children must learn to recognize during the very first expedition. Distinguishing between plants and fruits that are *edible, medicinal,* and *poisonous* is a very important and ongoing aspect of the learning experience in the forest.

Animals of the forest

Youngsters are thrilled by the sight of animals living freely in their forest home. Asking simple questions will stimulate children's curiosity and direct their observations. What are the colors in a butterfly? Can beetles fly? How do snails protect themselves? How do worms move? How many legs does a millipede have? Questions such as these can help children understand that millipedes are not worms even though they look like them, that snails and worms are also quite a different matter, and so on.

Children can also be taught that life in the forest is everywhere—under the moss, in the bark of the trees, etc. Depending upon the interests expressed by the children, subsequent outings could focus on more specialized topics. For example, a day could be spent looking for different kinds of birds. And, of course, this kind of concentration will lead to its own set of fascinating questions: How do birds build their nests? How do they fly? What do they eat? And so on.

The forest in danger

Our forests offer infinite possibilities for the education of the young child. But the forest is in danger from pollution, acid rain, indiscriminate cutting of trees, and forest fires. One of the most important missions of the teacher and parent should be to make the children aware, within the limits of their understanding, of the desperate need we all have to protect our forests from the carelessness, thoughtlessness, and selfishness of those who seem eager to destroy them.